YAS

ALLEN COUNTY PUBLIC LIBRARY

P9-BIV-014

THE LIBRARY OF
FUTURE ENERGY

OIL POWER OF THE FUTURE

NEW WAYS OF TURNING PETROLEUM INTO ENERGY

LINDA BICKERSTAFF

THE ROSEN PUBLISHING GROUP, INC.
NEW YORK

Published in 2003 by The Rosen Publishing Group, Inc.
29 East 21st Street, New York, NY 10010

First Edition

Library of Congress Cataloging-in-Publication Data

Bickerstaff, Linda.
Oil power of the future: new ways of turning petroleum into energy / by Linda Bickerstaff.— 1st ed.
p. cm. — (Library of future energy)
Summary: Presents the pros and cons of using oil for fuel as the growing demand for electricity increases problems such as air pollution.
Includes bibliographical references and index.
ISBN 0-8239-3662-7 (library binding)
1. Power resources—Environmental aspects—Juvenile literature. 2. Petroleum as fuel—Environmental aspects—Juvenile literature. [1. Fossil fuels—Environmental aspects. 2. Power resources.] I. Title. II. Series.
TD195.E49 B53 2002
333.793'2—dc21

2002002505

Manufactured in the United States of America

CONTENTS

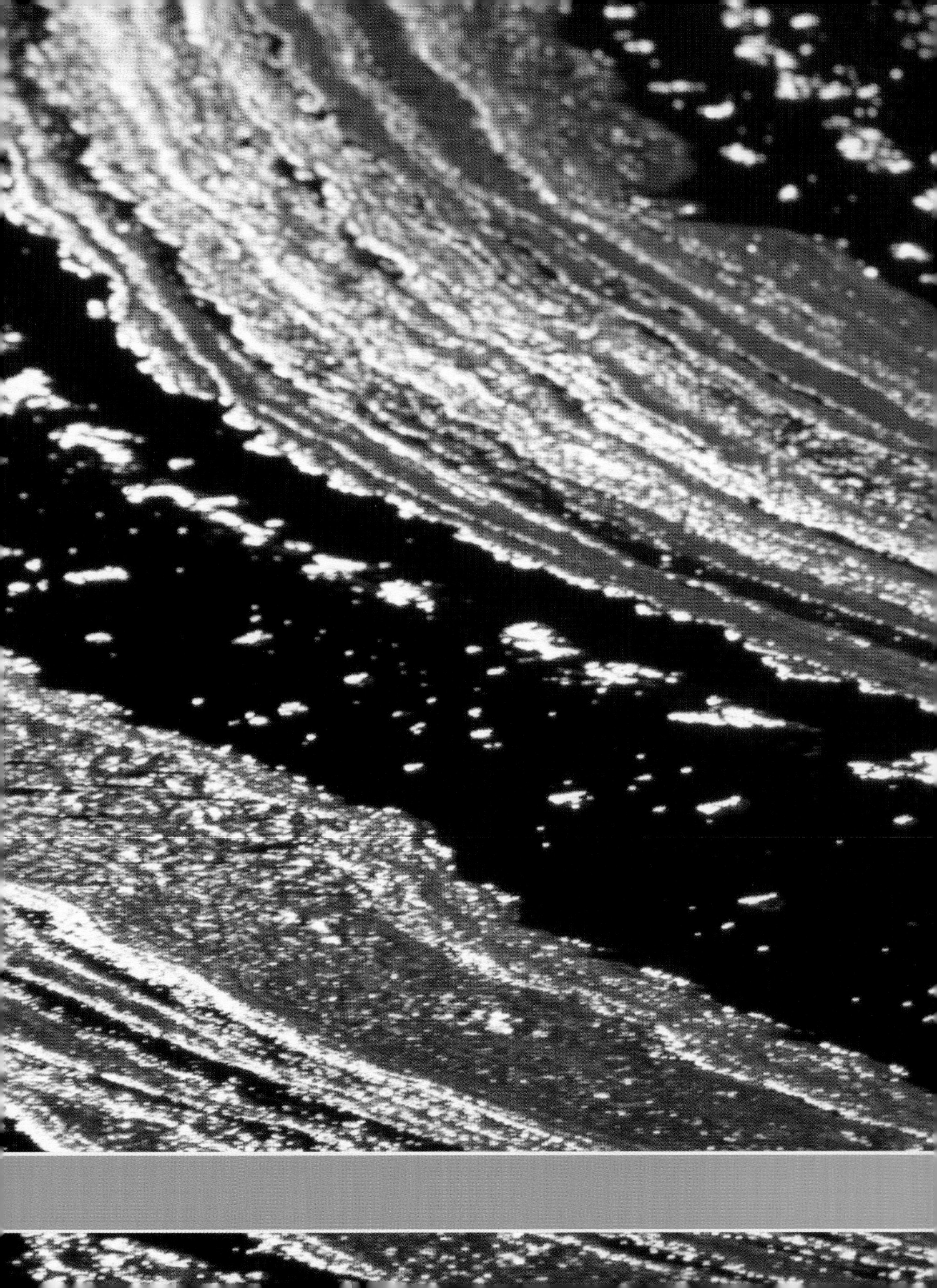

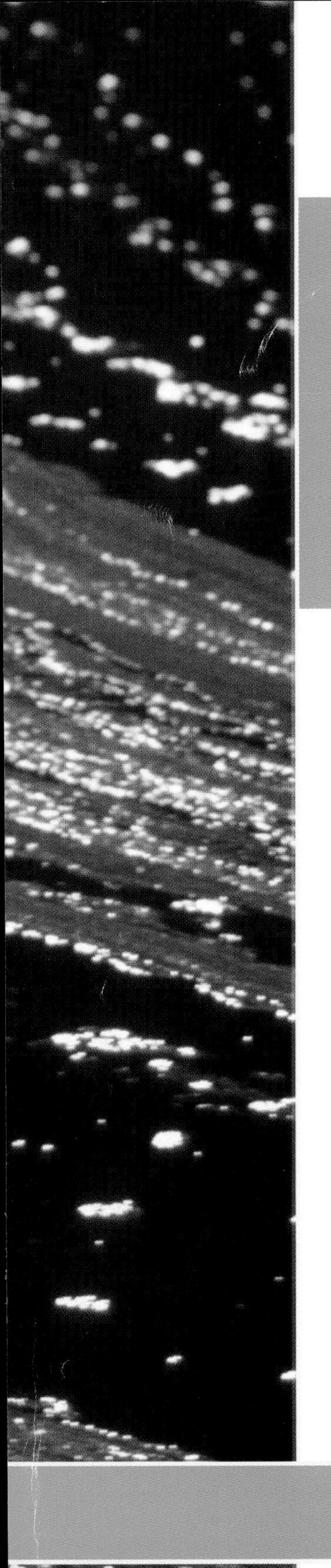

INTRODUCTION

Energy! The Greeks called it *energeia* from their word *energos*, meaning active. To us, energy means motion and power. Energy from the food we eat gives us pep, pizzazz, and get-up-and-go. Energy lights our towns, fuels our cars, cooks our pizza, and powers our computers. Energy sustains our lifestyles and life itself.

We are surrounded by energy. Sunlight, wind, ocean waves, and the flow of water in our rivers are all sources of energy. These are examples of kinetic energy, the energy of motion. Sources of kinetic energy are renewable. They can be used over and over again.

Another type of energy is potential, or stored, energy. It is bottled up in resources

Crude oil spurts from a well in Fremont County, Colorado. Like other fossil fuels, oil is made of decomposed plant and animal matter that has been buried deep in the ground for millions of years.

we call fuels. Potential energy must be changed to kinetic energy before we can use it.

The most popular fuel in the world today is fossil fuel. Formed millions of years ago deep in the earth, it has been used as an energy source for thousands of years. One of the fossil fuels is called petroleum, or oil. Providing almost half of all the energy used in the United States today, oil is equally important worldwide.

Oil and the other fossil fuels—coal and natural gas—are nonrenewable energy sources. Once they are gone, they are gone for good. The role that these fuels play in the future depends, in part, on how we use and conserve them today.

Oil Policies

Residents of San Antonio, Texas, stand in line to receive food through a relief program during the Great Depression, a period of severe economic downturn in the United States economy that lasted from 1929 to the early 1940s. Oil prices fell dramatically during this time.

For most of us, and for most nations of the world, energy means oil. National and international energy policies have dealt primarily with issues related to oil. Surprisingly, it wasn't the federal government that took the lead in forming the United States's first energy policies. It was the oil-producing states themselves.

For almost 175 years, the United States produced more oil than it used. Oil literally gushed out of the ground from hundreds of wells. People gave little thought to conserving it.

In 1934, with America deep in the Great Depression, the oil industry fell apart. Oil producers flooded the market with so much oil that its price fell to as little as 10 cents per barrel. America was wasting huge quantities of oil. Fearing that the federal government would take control of the oil industry to correct these problems, E.W. Marland, Governor of Oklahoma, invited representatives from other oil-producing states to join Oklahoma in forming a compact.

E.W. MARLAND: OIL MAN AND VISIONARY

E.W. Marland was born May 8, 1874, in Pittsburgh, Pennsylvania. Although trained in law, he was more interested in the oil industry. After making and losing his first million dollars in the Pennsylvania oil fields, he moved to Oklahoma in 1908. Three years later he struck oil on land leased from the Ponca Indians. Marland Oil Company flourished, and by 1922 it controlled one-tenth of the world's oil reserves. Falling victim to a hostile takeover by J.P. Morgan, Marland Oil Company was merged with a small Colorado company to become Conoco Oil. Marland, ousted from his own boardroom, was elected to the U.S. House of Representatives in 1932. In 1934, he was elected the tenth governor of Oklahoma. It was his vision and leadership that led to the development of the Interstate Oil and Gas Compact in 1935.

The immediate goal of this compact was to put a stop to overproduction and waste of this precious natural resource.

After several meetings, the Interstate Compact to Conserve Oil and Gas was born. Ratified by the legislatures of Colorado, Illinois, Kansas, New Mexico, Oklahoma, and Texas, it was approved by Congress on August 27, 1935.

The Interstate Oil and Gas Compact Commission (IOGCC) was established to carry out the programs of the compact. The first agency to develop energy policies in the United States, it survives to this day, representing governors of thirty-seven states. Its mission statement says, "The IOGCC is an organization of states which promotes conservation and efficient recovery of domestic oil and natural gas resources while protecting health, safety, and the environment."

Forty years passed before the United States government developed its first national energy policy. Before 1950, the U.S. produced more oil than any nation in the world. It also produced more than it used. In 1949, about 13 million barrels of oil per day were used in the U.S. Throughout the 1950s, U.S. oil production and consumption were about equal. By 1960, the U.S. could no longer produce as much oil as it needed so it began to buy oil from other countries—mainly from Saudi Arabia. By 1973, the United States was consuming 35 million barrels of oil a day and planned to import 12 billion barrels of oil from the Middle East. Oil producing countries of the Middle East imposed an embargo on oil shipments to the United States because America had supported Israel in the Yom Kippur War that Israel fought against Egypt and Syria. This embargo led to major gasoline and fuel-oil shortages in the United States and to a world wide recession, a period of reduced economic activity.

President Richard Nixon, in response to the embargo, established the first U.S. energy policy. He ordered the development of

Signs like this were common across the United States in the years following the 1973 oil embargo.

ways to conserve energy. Federal funding became available for research into alternative energy sources.

Since 1973, United States energy policies have reflected the availability and price of oil, and the attitudes of the political party in power. In general, when oil prices are high, policies support more conservation and increased emphasis on the development of other energy sources. When oil prices fall, we forget the lessons of the past. Conservation efforts slack off and the government slashes funds to develop alternative sources of energy.

Shortly after taking office in January 2001, President George W. Bush established a National Energy Policy Development Group made up of a panel of top government officials. Its purpose was to develop an energy plan for the future. A report from this group, issued on May 17, 2001, recommends that exploration for oil continue, and that new wells be drilled in the Arctic National Wildlife Refuge in Alaska. It also promotes the recovery of additional oil from existing wells, as well as spending more money to develop ways to harness renewable energy sources.

The report suggests that we take another look at nuclear power as a major source of energy. It also recommends many ways to increase conservation of our energy resources. We will have to wait to see how President Bush and his administration follow up on these recommendations. National, international, and environmental issues will decide the fate of oil as an energy source for the future.

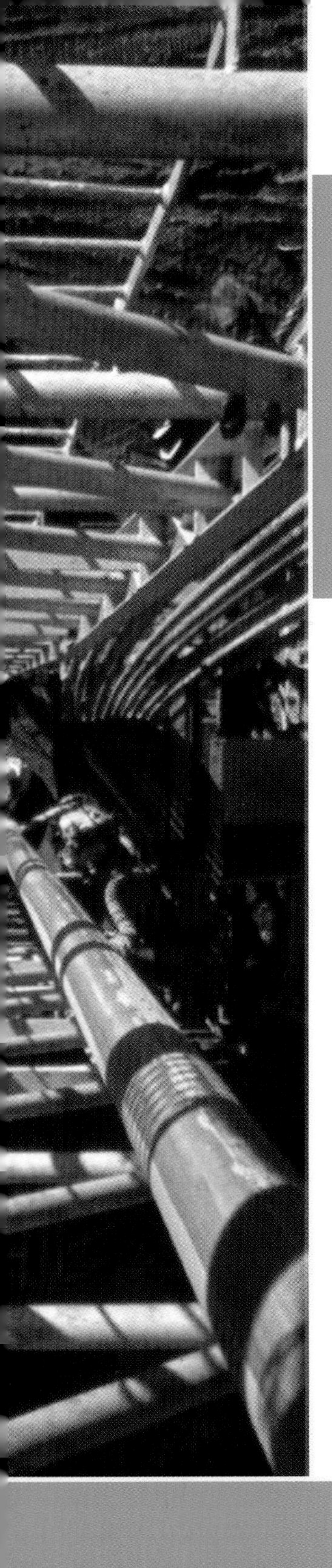

1 THE BASICS OF OIL

Oil, also called petroleum, formed in the earth from once living plants and animals. Gasoline is the most important product made from petroleum. One major oil company uses the phrase, "Put a Tiger in Your Tank" in its advertisements to suggest that its brand of gasoline gives your car more power and energy than other brands. But if you have a tiger in your tank, it's not the tiger from the ads. The story of how a tiger, a saber-toothed tiger, gets into your gas tank is the story of oil!

A Tiger Becomes Oil

When the earth was formed 4.5 billion years ago, it was extremely hot. Over millions of years, as its outer surface cooled, it

Exxon uses the tiger in its logo to suggest that its gasoline has extra energy.

developed an atmosphere containing oxygen, hydrogen, and other gases. Eventually oxygen and hydrogen combined to form water which fell to Earth as rain, establishing oceans and seas. It took 1.1 billion years of change before the earth could support life.

The oldest deposits of oil are 250 million years old. The oil found in the United States is much younger, formed 12 to 23 million years ago.

At that time, saber-toothed tigers and many other prehistoric animals roamed the shores of the shallow seas that covered much of what would one day become America. When a saber-toothed

TRUTH IN ADVERTISING

The name saber-toothed tiger is misleading because these frightening and ferocious predators of the Cenozoic period were not ancestors of present-day tigers. They are closely related to true cats, of the family Felidae and subfamily Machairodontinae. It is more appropriate to call them saber-toothed cats.

tiger died, its decomposing body washed into the sea, as did the remains of other animals and plants. This was organic matter.

In the sea, organic matter was attacked by algae and bacteria, simple one-celled plants and animals which broke down the complex structures, including the tiger's body, into their original building blocks, or elements—carbon, hydrogen, oxygen, and nitrogen. When the bacteria and algae died, they settled to the bottom of the seas along with fine particles of clay and sand called sediment.

Over millions of years, layers of sediment piled up. Under high pressure and heat, water was squeezed out, and the sediments began to turn into rock. Meanwhile, the organic material, including the remains of the tiger, changed into kerogen. Eventually, kerogen lost nitrogen and oxygen. The remaining hydrogen and carbon combined to form hydrocarbons—petroleum and natural gas.

The oil and gas oozed out of the sediments until they reached a layer of very hard rock. When the gas and oil could go no further, they filled pockets within the rocks or formed pools under rock shelves and salt domes. Here, they waited to be discovered.

Oil Is Discovered

In the early days of oil exploration, doodlebuggers, people who hunted for oil, looked for oil seeps and tar pits to give them clues about where to drill. They relied on intuition and guesswork to lead them to oil.

In this 1958 photo, an oil company worker plants geophones on a prospective drilling site to record the electrical impulses of the sound waves created by the mechanical thumpers.

By 1848, the search for oil started to become more scientific. Robert Mallet, an Irish geologist established the science of seismology, the study of motion under the earth. Mallet showed that sound waves, traveling through the earth, are reflected back to the surface when they hit rocks and other underground formations. The speed with which the waves are reflected and their angles of reflection depend on what they hit.

By 1919, seismology was used extensively in oil exploration. Early seismologists used dynamite explosions to create underground sound waves. Dynamite was dangerous and often unreliable, so mechanical "thumpers" were developed to replace dynamite as the sound source.

Today, the Vibroseis is used to generate continuous low-frequency sound waves instead of bursts of sound which were created with dynamite. Instruments called geophones pick up reflected sound waves and change them into electrical impulses which are recorded and processed by computers. Geophysicists are able to study the seismic reflection profiles on data printouts to determine which underground formations are likely to contain oil.

DRILLING FOR OIL

To remove oil from the ground, it must be mined or drilled. The first oil well in the United States was dug by hand. Workers hammered long pipes into the ground when oil seeped into the pits which had been dug.

Today miners use rotary drills made of very hard steel that spin or rotate drill bits deep into the earth. Rotary drills, like dentists' drills, have lubricants circulating through them to keep them cool and to flush out dirt. Two or three rotating cones, covered with sharp teeth, smash and grind rock into small fragments.

A geologist checks the debris in the well. If there are oil traces, the geologist removes the drill and lowers a testing tool into the hole to measure the pressure and volume of oil and gas in the well. The geologist can also estimate how much oil is in the hole. If the geologist decides there is enough oil to make the well a commercial success, the well is completed.

Pipes, called casing, are pushed into the wellbore around the drill. When the casing reaches a level just above the oil reservoir, it is cemented to the wellhead. The wellbore is drilled a little deeper, penetrating the reservoir.

A rotary drill bit viewed from within its shaft. Drill bits have complex heads made of multiple cones that have sharp teeth to penetrate the earth.

The flow of oil into the casing depends on the pressure within the oil reservoir, the amount of gas mixed with the oil, and the porosity of the rock of the reservoir. If the rock is very porous, oil can move easily out of the rock and into the casing. When the flow of oil from a well slows or stops completely, pumps are attached to move the rest of the oil up to the surface.

Oil on the Move

Oil is transported from wells to refineries through pipelines, on ships called crude carriers, or in large tanks that travel by trucks or trains. More than half of the oil is transported by ships called very large crude carriers (VLCC) and ultralarge crude carriers (ULCC). These ships are so large that there are only seventy ports in the world where they can dock. Most superports have offshore mooring areas and underwater pipelines to transfer oil from carriers to refineries. The only superports in the United States are off the coast of Louisiana.

OIL ON BOARD

- 1 barrel of crude contains 42 gallons
- 6 barrels of crude oil weigh 1 ton
- Very large crude carriers (VLCC) transport 200,000 to 300,000 tons of crude
- The minimum load of crude carried in a VLCC is 1.2 million barrels, or 50 million gallons
- Ultralarge crude carriers (ULCC) transport twice as much oil as VLCCs

Most of America's early oil wells were in Pennsylvania, so refineries were built nearby. Pipelines that carried oil to these refineries were relatively short. By 1914, however, most of America's oil came from midwestern and southwestern oil fields, especially those in Oklahoma and Texas. Oil was transported to East Coast refineries by a few small pipelines and by crude carriers from ports in Texas and Louisiana.

When America entered World War II, it was forced to change this system. A lot of oil was lost early in the war when German submarines sank crude carriers. As a result, the U.S. government joined with the oil industry to build a large-bore pipeline from the Midwest to eastern refineries. Called "the Big Inch," the technological and engineering skills used to build the pipeline have been surpassed only by those developed to construct the Trans-Alaska pipeline.

The *Arco Juneau* (above) was the first oil tanker to transport crude oil from Valdez, Alaska.

Oil Is Refined into Gasoline

Petroleum is called crude oil when it comes from the well. Crude oil, often called crude, varies from well to well.

Sometimes crude is heavy and sticky. Other crude is light and thin. It can be clear yellow to dense, opaque black. Some crude oil, called sour crude, contains a lot of sulfur so it smells like rotten eggs. Oil with little or no sulfur is called sweet crude. Light, sweet crude usually sells for higher prices on world markets than sour crude because it produces more gasoline when it is refined.

To refine, or distill, crude oil, it must be heated to a very high temperature until it turns into a vapor, just like water turns into steam. The components of crude oil vaporize at different temperatures. The lightest parts of the crude vaporize first. The heaviest parts vaporize last. As the vapor develops, it passes through a tall cooling tower. The first components to vaporize travel to the top of the tower because they are the lightest. The vapor from the heavier components stays at the bottom of the tower.

As they cool, the crude vapors condense, or turn back into liquid, at their respective sites in the cooling tower. They are now called distillates. The four products recovered by distillation are straight-run gasoline, naphtha, kerosene, and diesel fuel.

The portion of the crude left after distillation is called heavy bottoms. It can be refined into gas oil, lubricating oil, and asphalt by another refining process. Distillates can be purified even more through a process called "cracking," in which heat and pressure or a chemical catalyst "cracks" the heavier distillates into more valuable products like gasoline. The process used to refine each type of crude varies, depending on the properties of the crude itself and on what end-products are needed. For example, if gasoline is in short supply, the refining process can be customized to yield more gasoline. As winter approaches, less gasoline and more heating oil may be produced.

Putting the Tiger in Your Tank

Once gasoline has been refined, it is transported to your local service station, where it is pumped into underground holding tanks. You take the final step when you drive to the gas station, insert the nozzle, and press the appropriate levers to pump the tiger into your tank!

2 THE HISTORY AND DEVELOPMENT OF OIL

Archaeologists have found evidence that people used petroleum products as long as 6,000 years ago. Mosaic tiles discovered in ancient Babylonia were glued together with a thick, sticky form of oil called asphalt, which was also used by Egyptian morticians to soak the linens they used to wrap mummies.

The first known written records of petroleum come from Persia. Thousands of years ago, oil and gas seeps located in what is now Iran were ignited, perhaps by a stray spark or a bolt of lightening, sending flames shooting into the air. Thinking these natural torches were divine, people worshiped them.

Like the ancient Babylonians, Indians of Central America and Mexico used asphalt as

The Aztec Pyramid of the Sun at Teotihuacan, Mexico, is an example of the pyramids that the early Central Americans built using asphalt as cement.

cement for pyramids and on mosaics. The Seneca and Iroquois Indians used oil as medicine, paint, and for ceremonial fires. Indians in California used asphalt to waterproof baskets and boats.

The Modern History of Oil

In 1815, near Pittsburgh, Pennsylvania, shallow wells, which had been dug to collect brine with which to make salt, became contaminated with oil. Samuel Kiers, a salt maker, bottled some of the oil from these wells. He knew that the Seneca Indians used oil as medicine so he called his product Seneca Oil. He claimed that it could do everything from growing hair to curing the common cold. In truth, it did neither! Eventually the name Seneca Oil turned into Snake Oil, a term which today is used to indicate something that cannot do what it claims.

In 1849, Dr. Abraham Gresner, a Canadian geologist, distilled crude oil, producing kerosene. Benjamin Sillian, a chemistry professor at Yale University, reported in 1855 that oil could be used to make kerosene as well as paraffin and lubricating oil. The production of kerosene by these two men was important in the evolution of the oil industry. This development also had environmental significance.

Before the production of kerosene, whales were hunted and killed in huge numbers to provide oil for lamps. With the discovery of kerosene, the whale oil business died—just in time to save whales from extinction!

The Pennsylvania Rock Oil Company, one of the early suppliers of oil for kerosene distillation, obtained oil from huge, hand-dug pits, which were slow and expensive. Owners of the company decided to drill for oil the way people drilled for water. They hired a train conductor, Edwin L. Drake, to supervise drilling on their property near Titusville, Pennsylvania. Beginning in early 1859, after several months of drilling without striking oil, the scheme became known as "Drake's Folly." Finally, the investors sent a letter instructing Drake to give up. Fortunately for the oil industry, it took a long time for the letter to reach him. The day before the letter arrived, oil was found in the casing of the well and the American oil industry was born.

The Model-T and America's Oil Industry

In 1859, there was only one well in the United States. It produced 2,000 barrels of oil. One year later, 240 wells had been drilled. Together they produced 500,000 barrels of oil.

This photograph of the Drake Well, the world's first oil well, was taken in 1866.

From 1859 to 1900, the most important petroleum product worldwide was kerosene. People used it for light. Gasoline, a byproduct of kerosene distillation, was burned off because no one had any use for it.

In 1901, Henry Ford began mass production of the Model-T Ford. These cars needed large amounts of gasoline to fuel their engines, and suddenly there was a big demand for the previously unwanted gasoline. The automobile industry stimulated the growth of America's oil industry.

Oil exploration shifted from Pennsylvania to Oklahoma and Texas, and then to other southern and southwestern states. Oil was also discovered in California. Boomtowns sprang up overnight and forests of oil derricks dotted the hills. Oil derricks were everywhere, even on the lawn of the Oklahoma State Capitol.

By 1920, all the "easy oil" seemed to have been found. Fortunately, "Dad" Joiner, a wandering wildcatter, kept exploring. A wildcatter is a person who hunts for oil in places where it hasn't been found before.

In 1927, he leased 10,000 acres of land from Daisy Bradford, an East Texas rancher, and started drilling. In October, 1930, the largest oil field in the United States was discovered when Joiner struck oil at Daisy Bradford #3, the third well he drilled on the Bradford ranch. Eventually 25,000 wells were drilled on East Texas fields, some producing several thousand barrels of oil a day. It was the production from these fields that dropped the price of oil to 10 cents a barrel and resulted in the development of the Interstate Oil and Gas Compact Commission in 1935.

This is the oil derrick, nicknamed "Petunia," which stands on the grounds of the Oklahoma State Capitol. Plans are underway to transform the site into an exhibit.

War Leads to Worldwide Oil Exploration

Long before Edwin Drake drilled the first oil well in America, oil wells had been drilled in Russia, Indonesia, China, Poland, and

Canada. They supplied small amounts of oil for local use but little was done to develop large oil fields until World War II.

Before World War II, British warships were fueled by coal. British leaders knew that these ships would be at a disadvantage if they were used against the oil-fueled ships of Germany in another war. Since the British Isles had no significant oil deposits, the British government bought 51 percent interest in the Anglo-Persian Oil Company, which had been formed in 1909 by William Knox D'Arcy, a British oilman.

The outcome of World War II was heavily influenced by the fact that the world's oil supply was largely controlled by Britain, Russia, and the United States. Germany lost the war, at least in part, because it ran out of fuel.

After the war, Middle Eastern oil production began to rival that of the United States. Seven major international oil companies controlled the production of oil in the Middle East. Called the Seven Sisters, five of these companies were based in the United States: Gulf, Texaco, Mobil, SoCal, and Exxon. Royal Dutch/Shell was the sixth sister and British Petroleum the seventh. By the mid-1960s, Middle Eastern oil production surpassed U.S. production.

Petroleum in Alaska and Off-Shore

In 1968, the Atlantic Richfield Company (ARCO), drilling on the North Slope of Alaska, discovered the biggest oil field in North America.

It was not until 1977, however, that oil from this field became available. The problem was transportation.

To bring the Alaskan oil to market, it was necessary to build the Trans-Alaska Pipeline System (TAPS) to transport the oil from Prudhoe Bay to Valdez, where it could be loaded into crude carriers. The extremely cold temperatures in northern Alaska required the development of a new technology. Production from North Slope fields is beginning to drop, but the technology developed there will not go to waste. It is likely that more oil will be found in Alaska if wells are drilled in the Arctic National Wildlife Refuge as recently proposed in President Bush's new energy policy.

Shown here is a section of the 799-mile long Trans-Alaska pipeline that crosses over the Tanana River.

Off-shore drilling, while not a new concept, will lead to the discovery of many new oil fields under the oceans. Because of the danger and the huge expense involved in off-shore drilling, companies who drill these wells must be sure that their wells will yield enough oil to justify the expense. Most companies will not drill unless they think they will recover at least 300 million barrels of oil.

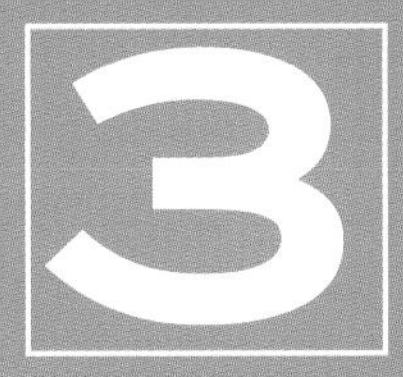

3 THE BUSINESS OF OIL

Ownership of oil in the United States is tied to ownership of land and mineral rights. For example, the Osage Indian tribe once owned most of the land in northeastern Oklahoma. Tribal elders distributed ownership of land parcels to tribe members, who were free to sell their allotments. The mineral rights to the land, however, were retained by the tribe.

In the early 1900's, oil was discovered on the Osage reservation. Many wildcatters leased land and drilled for oil. If they struck oil, seven-eighths of the money earned went to the wildcatter and one-eighth went to the tribe. The landowner did not receive any oil royalties unless he was a member of the tribe.

The United States government is the only government in the world that does not own the oil it produces. Oil companies do, however, lease government land and pay royalties to the federal government for the oil they drill on that land. Both federal and state taxes on the sale of gasoline and other petroleum products go to state and federal bank accounts.

The American Petroleum Institute points out that if a gallon of gasoline costs $1.00 at the pump, $0.43 of that price goes to pay federal and state taxes. Each gallon costs about fifty cents to produce and distribute, leaving seven cents a gallon profit to the oil industry.

United States oil companies are owned by thousands of shareholders. In most cases, many of the shares are owned by company employees who have obtained them through employee savings and retirement plans. According to Robert Anderson, author of *Fundamentals of the Petroleum Industry*, assets of the ten largest oil companies in the United States combined are in excess of $200 billion. Thousands of shareholders share this wealth.

In other countries, governments own at least some of the oil assets. Most governments take between 80 and 90 percent of the net gain from oil and gas. Oil reserves and production in certain countries are controlled by their governments. Oil resources in Mexico have been owned by the Mexican government since 1938.

America Depends on Oil!

In his book, *Energy: Shortage, Glut, or Enough*, Doug Dupler points out that the United States consumes the largest amount of oil in the world.

Each American uses the equivalent of seven gallons of gasoline a day. Home to only 5 percent of the world's population, the U. S. consumes 26 percent of the world's oil. According to *Energy Alternatives*, each American uses 24 barrels of oil per year. In Western Europe, 12 barrels of oil are used by each person every year. Africans who live below the Sahara Desert, some of the poorest people in the world, use less than one barrel of oil per person each year.

OPEC ministers begin a formal meeting at OPEC's headquarters in Vienna, Austria.

In 1998, the world's oil consumption was 73.6 million barrels a day. Americans consumed 18.9 million barrels a day, while Japan, China, Russia, and Germany combined consumed just 15 million barrels a day!

In 1960, when the United States began to import oil, businessmen and government officials from other oil producing countries formed the Organization of Petroleum Exporting Countries (OPEC). Today member countries include: Algeria, Indonesia, Iran, Iraq, Kuwait, Libya, Nigeria, Qatar, Saudi Arabia, United Arab Emirates, and

Venezuela. With about 52 percent of the world's oil reserves, and 6 percent of the world's population, these countries together consume only three percent of the world's energy.

How Oil Affects America's Economy

Since member nations of OPEC consume so little of it, they have a lot of oil to export. If each country in OPEC exported as much as possible, the world would be flooded with oil and the world price of oil would drop. By agreeing on how much oil each country exports, OPEC plays a major part in controlling the price of oil.

For oil importing nations like the United States, a world oil price of about $25 per barrel seems to be the "break even" price. When OPEC exports more oil, causing world oil prices drop below $25 per barrel, importing nations have no incentive to produce more oil themselves because production costs are higher than the purchase price. But when OPEC decreased oil exports in 1981, the price of a barrel of oil increased to $56.50 per barrel. The United States looked for ways to recover more domestic oil and emphasized oil conservation. When oil prices dropped again, these efforts were abandoned.

Lessons from the past should teach us that no matter what the cost of oil, we should continue to develop alternative sources of energy and always practice oil conservation. The U. S. economy is hurt when we buy more goods than we sell, making us a debtor nation. By importing less oil and by increasing the production of goods to export, our balance of payments would improve.

Oil doesn't just fill your gas tank. There are thousands of substances derived from oil which are used to produce everyday items like paint, tires, and even synthetic fabrics.

Petrochemical products made from oil, such as plastics, tires, fertilizers, paints, carpeting, and medications are just a some of our exports to other countries. Carbon fiber, recently added to America's petrochemical industry, may become one of our biggest exports. Made from the heavy bottoms left after gasoline is refined from crude oil, it is extremely light and strong. Carbon fiber support beams are being used in place of steel frames in the construction industry. Many of today's new cars contain parts made from carbon fiber. One young entrepreneur in Minnesota is using it to make kite struts, violin bows, and kayak paddles. The sale of these and other petrochemical products will begin to correct the balance-of-payments problem.

4 THE PROS AND CONS OF OIL AS AN ENERGY SOURCE

No energy source is perfect. In considering energy sources, we must weigh the advantages against the disadvantages.

OIL AS A SOURCE OF ENERGY

Oil is a good source of energy because it is efficient and cheap. The British thermal unit (Btu) is used to compare the efficiency of different types of fuel. As reported by Robert R. Wheeler and Maurine Whited, authors of *Oil: from Prospect to Pipeline,* a cord of wood produces 10.4 million Btus of heat while a ton of coal produces 22.4 million Btus. The refined products of one barrel of oil collectively produce 38.6 million Btus.

Oil is the most efficient of these fuels. Only natural gas and nuclear energy can rival it for efficiency. Oil is also cheap when compared to other fuels. OPEC nations are able to produce oil for about $2 a barrel. Oil production in the United States is more expensive, at about $14 per barrel. We import oil for about $25 a barrel. A cord of wood is not only a less efficient fuel but it costs a lot more—as much as $120. To date, the efficiency and cheapness of oil as a fuel have outweighed the many negative aspects of its use.

Using Up Oil Supplies

Oil is a nonrenewable resource that we will someday use up. The United States Geological Survey (USGS), in its March 2000 report, estimated that there are 646 billion barrels of oil yet to be discovered in the world, outside of the United States. Most of this oil is in the Middle East and in the offshore areas of western Africa and eastern South America. The report also suggests that existing, nonproducing oil fields still contain over 600 billion barrels of oil, much of which can be recovered by new techniques.

Other people disagree with this optimistic assessment by the USGS. In the recently published *Hubbert's Peak: The Impending World Oil Shortage*, Kenneth S. Deffeyes, a petroleum engineer and professor at Princeton University, says that global oil production will reach its highest level sometime between 2004 and 2008 and will never rise again. Deffeyes feels that we must move as quickly as possible to alternative fuels. As he says, "Fossil fuels are a one-time gift that lifted us up from subsistence agriculture and eventually should lead us to a future based on renewable resources."

The Hidden Costs of Oil

Reliance on oil as our primary source of energy is dangerous and has many hidden costs. For instance, we may use up oil supplies before we have developed good alternative energy sources. Use of oil contributes to global warming and the development of acid rain. Air pollution from the use of oil damages and destroys the ozone layer, increasing the incidence of skin cancers. Oil spills pollute our land and water, killing plants and animals that help maintain the earth's balance.

When the Temperature Rises

Global warming is real. When the sun strikes the earth, kinetic energy from sunlight is converted into heat. Some of the heat is absorbed by oceans and landmasses. The rest radiates into the earth's atmosphere. Atmospheric carbon dioxide (CO_2) and methane, along with other gases, trap some of this heat just as the glass in a greenhouse traps heat. This phenomenon is called the greenhouse effect. Natural greenhouse effect is necessary to maintain life on earth. Cornelia Blair, in her book, *The Environment: A Revolution in Attitudes*, points out that without natural greenhouse gases in the atmosphere, the earth's surface would be at least thirty-three degrees colder than it is, making the earth uninhabitable.

The oil products that we burn for fuel release carbon dioxide into the air. Carbon dioxide spews from the tailpipes of our cars, the smokestacks of our factories, and the chimneys of our homes. It rises into the earth's atmosphere. There it is joined by gases emitted from the burning of rain forests in South America, the burning of slash from the clear-cut forests of the Pacific Northwest, and from the smoldering fires of thousands of trash dumps around the world. Over the last 150 years, the amount of CO_2 in our atmosphere has increased by over 25 percent and has markedly intensified the greenhouse effect.

Pollution from automobiles is one of the biggest contributors to urban smog, the greenhouse effect, and a variety of breathing problems.

According to Blair, since the measurement of global surface temperatures began in the mid-1800s, the temperature of the earth's surface has increased about one degree Fahrenheit. It is projected to increase by two to six degrees over the next hundred years.

Scientists today can see the early effects of global warming. Glaciers have begun to melt more rapidly, causing a rise in sea levels throughout the world. In the last century, sea levels have risen ten inches. They are predicted to rise another eight inches by 2030, and twenty-six inches by the year 2100. Low-lying land along coastlines will eventually be covered with water. Wetlands, river deltas, and many populated areas along coasts will be radically changed, if not totally obliterated. Global warming is also altering weather patterns. We are seeing increases in extreme weather events such as tornadoes, hurricanes, and typhoons.

Global warming is raising sea levels, endangering low-lying coastal areas. Scientists estimate that Martha's Vineyard *(top)* could be flooded in the next 100 years. The lower picture is a computer-generated image of what the area would look like if that happened.

Acid Rain

Another negative effect of using oil is the formation of acid rain. Acids are water-soluble chemical compounds that taste sour, sting your eyes, and make your skin burn. Sulfur dioxide and nitrogen dioxide are byproducts of the combustion of oil products. When they are dissolved in water, these gases form sulfuric acid and nitric acid. After they form in the atmosphere, they fall back to Earth as acid rain. The leaves of trees and other plants are burned by acid rain. Plants grow more slowly; some die.

Acid rain also changes the chemical nature of lakes and streams. When water becomes too acidic, fish eggs can't hatch. The small plants and animals on which fish feed die. Many species of fish may be lost if this trend continues. Perhaps the worst effect of acid rain is that it washes valuable minerals out of the soil.

Oil Spills

Oil spills can happen anywhere—at wellheads, from drilling rigs, crude carriers, pipelines, and storage tanks. The National Academy of Science estimates that two million tons of oil enter the oceans every year during the everyday transportation of oil. Small amounts of oil may be spilled when tankers connect or disconnect from underground pipelines. A bit more enters the ocean when it is not completely burned in the engines of crude carriers or other ships. Eventually all of these small spills add up to a lot of oil—as much as twenty times more oil than is spilled in accidents.

An amazing amount of oil is deliberately dumped into oceans, rivers, and lakes. People on about half of the world's merchant ships and many pleasure boat owners dump oily waste from their tanks

Since oil spills endanger marine life, oil companies are now required to maintain ships and equipment capable of cleaning spills. The ship above was used for cleanup operations in the Gulf of Mexico in 1998 after a pipe from a British Petroleum drilling platform began to leak.

into the water. They don't think about the pollution they are causing. In addition, crews of many tankers clean their tanks between cargoes, dumping old oil out with the wash water.

The *Oil Spill Intelligence Report*, published by the Cutter Information Service, points out that the biggest oil spill in history occurred during the 1991 Persian Gulf war, when about 240 million gallons spilled from oil terminals and tankers off the coast of Saudi Arabia. The second largest spill occurred June 3, 1979, when 140 million gallons were lost at the Ixtoc exploration well in the Gulf of Mexico. The *Exxon Valdez* accident in March 1989 spilled

These penguins try to clean oil from their soaked feathers after being caught in a year-old oil slick off the South African coast.

approximately eleven million gallons of crude into Prince William Sound, Alaska. Although it was the thirty-fifth largest spill in the world, it was the largest spill in the United States.

The impact of oil spills on oceans and shorelines may be catastrophic. Oil is poisonous to tiny marine creatures that supply much of the food for fish and marine mammals. As little as one gallon of oil in one million gallons of water is harmful to these tiny animals.

Marine birds and mammals are affected in many ways by oil spills. Birds are protected from becoming "waterlogged" by an intricate system of overlapping feathers that are connected to each other by tiny barbs. Outer feathers prevent water from reaching their underlying downy feathers and skin. When a bird becomes coated with oil, its feathers stick together, destroying its waterproofing system. Cold water quickly soaks into the insulating down and reaches its skin. Over time, the bird uses up its fat reserves trying to maintain its body temperature. It is at great risk of dying from hypothermia. In an attempt to rid itself of the oil, the bird's immediate response is to clean itself. As it nibbles at its feathers, it swallows a lot of poisonous material that damages its liver, lungs, kidneys, and other organs. Linda Schwartz, in *Earth Book for Children*, reports that the oil spill from the *Exxon Valdez* was responsible for the deaths of more than 30,000 seabirds, 980 sea otters, and 136 bald eagles.

Not all oil spills occur at sea. People dump used oil onto the ground or into sewer storm drains. Rain carries the oil into streams and creeks. The same thing happens when oil is spilled at wellheads, around refineries, or at gas stations. Natural seepages also occur and contribute to contamination of groundwater. Added together, these kinds of spills are as disastrous as a spill from a crude carrier.

The Dangers of Air Pollution

Air pollution, like water pollution, can seriously injure our health. The white haze that hangs over many cities is tropospheric ozone, or smog. This gas forms when some of the hydrocarbons and nitrogen oxides in car and truck exhaust fumes react together in the presence of heat and sunlight. The smog that forms can irritate the tissues of our lungs and can cause bronchitis and pneumonia. The added stress from smog can be fatal to people who already have heart and lung disease.

Atmospheric ozone is a colorless gas (O_3) that is normally found in the earth's atmosphere at altitudes below thirty miles. Atmospheric ozone protects the earth from most of the ultraviolet light that comes from the sun. Known to damage skin, ultraviolet light is the direct cause of several types of skin cancer. Chlorofluorocarbons (CFCs), which are petrochemicals used in air conditioners, refrigerators, and in pressurized spray-can products, are also contributing to the loss of atmospheric ozone. When chlorine is released from CFCs in large quantities, it causes holes in the ozone layer, allowing more UV light to reach the earth.

Over the last several years, the number of people developing skin cancers, especially potentially fatal malignant melanomas, has increased drastically. Ophthalmologists are also seeing more patients with eye problems, including blindness, which are directly related to exposure to UV irradiation.

What We Can Do

Attempts are being made through international treaties to address the dangerous effects of the burning of fossil fuels on the environment. Environmental groups and world governments are working with oil companies, auto makers, and other businesses to encourage voluntary reduction in greenhouse gas emissions and the banning of CFCs.

THE KYOTO PROTOCOL

In December 1977, a United Nations conference brought representatives of 160 nations to Kyoto, Japan, specifically to address problems of global climate change. The Kyoto Protocol, an international treaty, was the result. It requires that developed nations of the world reduce emissions of greenhouse gases to below 1990 levels by the year 2012. By March 2002, according to the official UN site, eighty-four nations had signed the treaty and fifty nations had ratified it. Fifty-five nations must ratify it before it becomes binding. To date, the U.S. Senate has not ratified the treaty.

Individually, each of us can figure out ways to use less fossil fuel. For a short trip, we could ride a bike or walk instead of taking the car. We can turn off the lights when we leave a room.

We can also think before we pollute. When we change the oil in our cars, we can dispose of it properly, instead of dumping it in the sewer. We can listen to what candidates for public office say about energy issues. We can encourage others to vote for people with sound energy policies who are willing to work to correct fuel-related environmental problems.

INTO THE FUTURE

Oil is today's major source of energy and will probably remain so for at least another ten to twenty years. How long oil retains its number-one status depends on many factors. Can new oil fields can be found? Can we develop new drilling techniques to reach previously inaccessible oil? Can we recover more oil from existing fields? How quickly and actively will manufacturers look for ways to improve fuel efficiency? What are the nations of the world, especially the United States, willing to do to protect the environment from the harmful effects of fossil fuels?

FINDING MORE OIL

Technology developed for space exploration is being used to help in the search

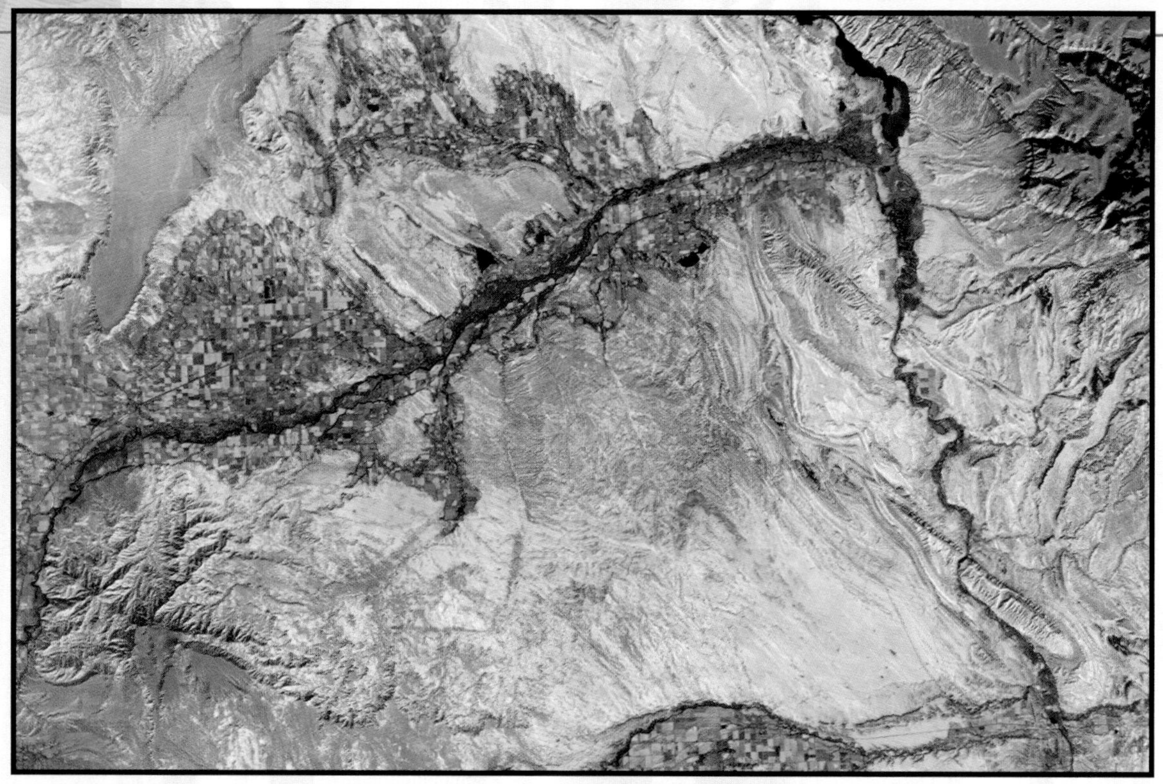

LANDSAT imaging has applications in petroleum and mineral exploration as well as in map-making, the study of water bodies, agriculture, and environmental sciences. The above image, which highlights rock formations and land use in Wyoming's Bighorn Basin, was taken from a LANDSAT satellite in 1991.

for new sources of oil. Aerial photography, which looks for appropriate geologic formations, has been used for many years. Much more sophisticated pictures of the earth's surface are now being obtained using LANDSAT satellites. Side-looking airborne radar (SLAR) is another tool being used. It has the advantage of taking excellent pictures regardless of weather conditions.

One hundred companies that specialize in oil and mineral exploration have joined together to form the Geosat committee. This group has been instrumental in developing a $100 million vehicle, the Stereosat, that takes extremely high-resolution stereo pictures of the earth's surface.

New techniques are also being developed for use on the ground. British Petroleum, ConocoPhillips Petroleum, and Amoco have hired scientists at Lawrence Berkeley Laboratories to develop new techniques to look at underground geology. The goal is to develop a model of subterranean rocks, faults, and oil pools of a proposed oil field. The model will allow drilling techniques to be individualized to improve the chances of striking oil. These techniques, once they are perfected, will also help people decide the best way to recover additional oil from existing wells.

NEW DRILLING TECHNIQUES

The United States Department of Energy (DOE) is one of the leaders in the development of new drilling techniques. The DOE is investing in research to develop "smart wells." "Zero footprint" drilling techniques, rig-less drilling, and self-drilling well technology are three of the areas that will be studied. The goal is to develop drilling technologies that will do the least amount of damage to the environment. Engineers are working to develop safety measures that can be conducted from an office rather than at a well site. One goal is to create a system that can anticipate drilling problems and correct them on the spot. Drilling techniques of the future may also use high-tech lasers.

Getting More Oil from Old Wells

Oil experts estimate that nonproducing oil fields still contain as much oil as has already been removed from them. Water flooding is the most important technique for recovering more oil from wells. Pumps force water, under high pressure, through an injection well next to a pumper well. The water pushes into cracks and fissures in the rock and forces oil to the pumper well. A similar technique uses steam instead of water to push the oil to the pumper well. First-time and second-time recovery by either of these methods can result in recovering about 50–60 percent of all the well's oil. Other recovery methods introduce detergents and chemicals called surfactants into wells to wash oil out of reservoir rocks.

Another technique uses carbon dioxide to flush the oil out of wells. In his book, *Fundamentals of the Petroleum Industry*, Robert Anderson says that Atlantic-Richfield Oil Company is piping CO_2 from gas wells in Colorado to West Texas. Using 20 trillion cubic feet of CO_2 at a pressure of 2,000 pounds per square inch, they plan to recover three to five billion barrels of reserve oil from existing oil fields.

Recovering oil with these new methods is very expensive. Oil prices will have to be quite high before it will be economically sound to use these techniques.

CONSERVING OIL

The Corporate Average Fuel Efficiency Act of 1975 required car manufacturers to increase mileage of new cars to 27.5 miles per gallon (mpg) by 1985. New cars in 1999 traveled an average of 28 mpg. Improved auto efficiency, however, has been offset by increased weight and power in new vehicles that consume more gas than ever before. In 1999, sales of sport utility vehicles (SUVs) and pickup trucks made up 46 percent of the U.S. auto market. SUVs and trucks do not have efficiency requirements and have less stringent emission requirements than cars. The 175 million cars and trucks on U.S. roads use more than 300 million gallons of gas each day.

As gas-guzzling sports utility vehicles get bigger and more fuel inefficient, some lawmakers are considering raising the sales tax on these cars. They hope the extra tax will deter some people from buying SUVs or at least help pay for some of the environmental damage they cause.

Greenpeace, an environmental activist group, estimates that if fuel efficiency for cars, SUVs, and trucks were increased to 32 mpg, the amount of fuel saved would equal oil imports from all Middle

REDUCING DANGEROUS EMISSIONS

Hybrid automobiles use both gasoline and electrical power for energy. The Honda Insight was constructed to get maximum miles per gallon of gas. It is very light, weighing only 1,800 pounds. It has a one-liter, three-cylinder gas engine, and is very aerodynamic. Its EPA (Environmental Protection Agency) mileage rating is 61 mpg in the city and 68 mpg on the highway. When accelerating and cruising, its gas engine does the work. Its electric motor provides the extra power it needs to climb hills and accelerate rapidly. When it slows down, the electric motor acts as a generator and recharges its batteries.

The Toyota Prius was built to minimize emissions in city driving. It is a bigger car than the Insight, weighing 2,785 pounds. Its electric motor starts the car and is the main energy source until it reaches 15 mph. At that speed, the gasoline engine is turned on. For highway driving, power is supplied by both the gas engine and the electric motor. When braking, the electric motor recharges the batteries. The Prius's EPA ratings are lower than those of the Insight, but the Prius still gets 45 to 50 mpg.

Eastern countries. Jack Riggs, staff director of the United States House of Representative's subcommittee on Energy and Power, says that if fuel efficiency increased to 40 mpg, oil consumption would decrease by 2.8 million barrels per day.

Hybrid automobiles are one answer to this problem. Honda was the first to develop a hybrid car called the Insight. It gets 61 mpg when driven in the city, and 68 mpg on the highway. Toyota is mass-producing a hybrid car called the Prius. This car, also fuel-efficient, was built primarily to reduce greenhouse gas emissions. Toyota is developing several other innovative vehicles in conjunction with General Motors and Volkswagen.

Showing the World We Care

The United States, as the largest consumer of oil in the world, must lead the way in recognizing and correcting environmental problems caused by using fossil fuels. If we don't set an example by decreasing emissions of greenhouse gases, regulating the use of petrochemicals that are destroying the ozone layer, and applying our considerable knowledge and technology to decreasing acid rain, how can we expect others to do so? We are the greatest offenders, therefore we must be the most concerned. We must work to conserve energy and develop renewable alternative fuels that are environmentally friendly. If we are successful, we'll have a safe, pleasant environment in which to live and we'll have enough oil for many years to come.

GLOSSARY

assets The property or resources of a business.

boiling point The point at which a fluid turns to vapor.

bring in To strike oil.

british thermal unit (Btu) The amount of heat required to raise the temperature of one pound of water from 62°F to 63°F.

byproduct Something extra produced in the process of making another thing; a secondary result.

catalyst A substance which either speeds up or slows down a chemical reaction without itself being changed.

combustible Capable of igniting into fire; flammable.

compact An agreement to work together to reach a goal.

derrick A framework which supports the mechanism for drilling oil.

doodlebugger A person who looks for oil; oil prospector.

evolution A process of development, growth, or change.

fossil fuels Materials formed in prehistoric times from the remains of plants and animals: coal, natural gas, and oil.

geologist A scientist who studies the structure of the earth's crust and its layers, rock types, and fossils.

hydrocarbon A compound made up of hydrogen and carbon.

lubricant A substance such as grease used to reduce friction.

mineral rights Ownership of minerals in a tract of land.

mosaic Bits of stone or glass set in asphalt or concrete.

petrochemical A product made of petroleum or natural gas.

pitch A sticky black substance formed when oil is heated.

policy A plan for goals and procedures of a government.

reservoir A place where liquids or gases are collected and stored in large quantities.

seep A place where petroleum oozes from the ground.

seismology The science of earthquakes; also concerned with movement of sound waves through the earth.

show Indication of the presence of metal, coal, oil, and other substances in the earth.

wellbore The center of the well which is being drilled.

wellhead The top of the well to which well casings are attached and to which gauges, pumps, and other equipment can be attached.

wildcatter A person who drills for oil in new areas.

FOR MORE INFORMATION

Alliance to Save Energy
1200 18th Street NW, Suite 900
Washington, DC 20036
(202) 857-0666

American Petroleum Institute
1220 L Street NW
Washington, DC 20005-4070
(202) 682-8000
Web site: http://www.api.org

Canadian Society of Environmental Biologists
PO Box 962, Station F
Toronto, ON M4Y 2N9
Web site: http://www.freenet.edmonton.ab.ca/cseb

Environment Defense National Headquarters
257 Park Avenue South
New York, NY 10010
(212) 505-2100

The Union of Concerned Scientists
2 Brattle Square
Cambridge, MA 02238
(617) 547-5552
Web site: http://www.ucsusa.org
e-mail: ucs@ucsusa.org

United States Department of Energy
1000 Independence Avenue SW
Washington, DC 20585
(202) 586-6503
Web site: http://www.fe.doe.gov

United States Geologic Survey
United States Department of Interior
119 National Center
Reston, VA 20192
Web site: http://www.usgs.gov/index.html

World Watch Institute
1776 Massachusetts Avenue NW
Washington, DC 20036-1904
(202) 452-1999
Web site: http://www.worldwatch.org

Web Sites

Due to the changing nature of Internet links, the Rosen Publishing Group, Inc., has developed an online list of Web sites related to the subject of this book. This site is updated regularly. Please use this link to access the list:

http://www.rosenlinks.com/lfe/oil/

FOR FURTHER READING

Anderson, Madelyn Klein. *Oil Spills*. Franklin Watts. A First Book. New York: Justin Books, Inc., 1990.

Beres, Samantha. *101 Things Every Kid Should Know About Science.* Los Angeles, CA: Lowell House Juvenile, 1998.

Gans, R. *Let's-Read-and-Find-Out Science Books: Oil the Buried Treasure.* Toronto, ON: Fitzhenry and Whiteside, 1975.

Schwartz, Linda. *Earth Book for Kids*. Santa Barbara, CA: The Learning Works, 1990.

Snow, T. *Global Change*. Chicago, IL: Children's Press, 1990.

BIBLIOGRAPHY

Anderson, R. *Fundamentals of the Petroleum Industry*. Norman, OK: University of Oklahoma Press, 1984.

Blair, C. *The Environment: A Revolution in Attitudes*. Farmington Hills, MI: Gale Group, 2001.

Deffeyes, K. *Hubbert's Peak: The Impending World Oil Shortage*. Princeton, NJ: Princeton University Press, 2001.

Dupler, D. *Energy: Shortage, Glut, or Enough?* Farmington Hills, MI: Gale Group, 2001.

Wheeler, Robert.and Maurine Whited. *Oil: From Prospect to Pipeline*. Houston, TX: Gulf Publishing Company, 1981.

INDEX

CREDITS

About the Author

Linda Bickerstaff, M.D., a retired general and peripheral vascular surgeon, grew up in Ponca City, Oklahoma, and is the daughter of a Conoco Oil Company doodlebugger. After retiring in 1995, she returned to Ponca and now works as a volunteer at "The Castle on the Prairie," the historic fifty-five-room mansion of Oklahoma oilman E. W. Marland.

Photo Credits

Cover © Bill Ross/Corbis; pp. 4–5 © Natalie Forbes/Corbis; p. 6 © Lowell Georgia/Corbis; p. 7 © Corbis; p. 8 © The Marland Estate; p. 10 © Bettmann/Corbis; pp. 12–13 © Jack Fields/Corbis; p. 14 © Reuters/Larry Downing/TimePix; p. 16 © Hulton/Archive/Getty Images; p. 18 © Jack Fields/Photo Researchers, Inc.; p. 20 © Paul A. Souders/Corbis; pp. 22–23, 27, 29, 41, 43, 44, 53 © AP/Wide World Photo; p. 24 © Danny Lehman/Corbis; p. 26 © from Pennsylvania Historical & Museum Commission, Drake Well Museum Collection, Titusville, PA; pp. 30–31 © David and Peter Turnley/Corbis; p. 33 © AFP/Corbis; p. 35 (left) Charles O'Rear/Corbis; p. 35 (right) Craig Aurness/Corbis; pp. 36–37 © Chromosohm/Photo Researchers, Inc.; p. 40 © Ecoscene/Corbis; pp. 48–49 © Kit Kittle/Corbis; p. 50 © Earth Sattellite Corporation/Photo Researcers, Inc.; p. 54 © Reuters NewMedia, Inc./Corbis.

Editor

Jill Jarnow

Design and Layout

Thomas Forget